KV-371-702

The Secret Life of
BUGS

happy yak

© 2024 Quarto Publishing plc
Text © 2024 Moira Butterfield
Illustration © 2024 Vivian Mineker Chen

Moira Butterfield has asserted her right
to be identified as the author of this work.
Vivian Mineker Chen has asserted her right
to be identified as the illustrator of this work.

Senior Designer: Sarah Chapman-Suire
Commissioning Editor: Carly Madden
Editor: Victoria Garrard
Consultant: Michael Bright
Creative Director: Malena Stojić
Associate Publisher: Rhiannon Findlay
Production Manager: Nikki Ingram

First published in 2024 by Happy Yak,
an imprint of The Quarto Group.
1 Triptych Place, London
SE1 9SH, United Kingdom.
T (0)20 7700 6700 F (0)20 7700 8066
www.quarto.com

No part of this publication may be reproduced, stored in a
retrieval system, or transmitted in any form, or by any means,
electrical, mechanical, photocopying, recording or otherwise,
without the prior written permission of the publisher or a
licence permitting restricted copying. In the United Kingdom
Such licences are issued by the Copyright Licensing Agency,
5th Floor, Shackleton House, 4 Battle Bridge Lane, London SE1 2HX.

All rights reserved.

A catalogue record for this book is available from the British Library.

ISBN 978 0 7112 8654 2

Manufactured in Guangdong, China TT102023
9 8 7 6 5 4 3 2 1

FSC
www.fsc.org

MIX
Paper | Supporting
responsible forestry
FSC® C016973

CONTENTS

I'm like a shiny bead with spots,
sitting on your flowerpots.
I fly around on tiny wings
and love the warmth that sunshine brings.
I'm Luna the ladybird. Hello!
I'll help you meet some bugs. Let's go!

Dear Reader,

I'm here to take you on a journey through the world of insects (we are often nicknamed bugs, too). Together we'll find out about the biggest and tiniest insects, the nighttime insects, and the noisy insects. We'll be discovering the cleverest builders, swimmers and flyers.

I'll be telling you some stories from long ago and from around the world, too. Butterflies, grasshoppers, bees and ants are all waiting to share their tales with you.

Oh, and look out for my teeny-tiny bug facts along the way.

Are you ready? OK! Let's see what we can find amongst the flowers and leaves.

Luna the ladybird

WHEN I WAS BORN

Tiny eggs and big changes

At the very beginning of summer last year, I was just a tiny egg. I was laid along with a cluster of other eggs on the underside of a leaf.

My egg was yellow and lemon-shaped.

I grew to about 1 cm long.

1 cm

An aphid. Yum!

After a few days I hatched out, but I didn't look like a ladybird yet. I was black and spiky with orange stripes – a larva! Most insects are larva when they first hatch, though they might look different to me. When I was a larva, I spent my time eating all the plant-eating bugs – called aphids – that I could find.

I grew quickly because of all those aphid snacks. A couple of times I got so big that my skin grew too tight, so I shed it and grew a new one. After about 3 weeks I glued myself to a leaf and turned into a little spotted blob called a pupa.

Pupa

Inside my pupa an incredible thing happened. My body changed completely! It's called metamorphosis (met-a-more-foh-sis). It happens to lots of insects. After a week or so I emerged as a pale-coloured ladybird, but I soon turned red and my black spots appeared.

I'm me at last!

TEENY-TINY BUG FACT
There are more than 5,000 different types of ladybird in the world. We are lots of different colours and we have different numbers of spots.

Ladybirds are a type of beetle. I am a seven-spotted ladybird.

ALL ABOUT ME
(AND MY INSECT FRIENDS)

Legs, wings and other things

Insects come in lots of different shapes and sizes,
but we all have some body parts that are similar.

Tiny claws

Mouthparts – I use
mine to chew up aphids.

Two **antennae** that help
us to smell, taste and
feel our way around.

I have a protective shield
behind my tiny head.
It's called a **pronotum**
(proh-no-tum). Not all
insects have one of these.

Eyes – see p30 for
more about our
amazing eyes.

Six **legs** with joints in them
(a joint is a bendy part,
like your knee joint).

A hard outer case
to protect the body.
It's called an **exoskeleton**.

A body in three parts – a head, a thorax (the middle part)
and an abdomen (the bottom part). My body parts are
usually hidden but you can see them more clearly on an ant.

Most, though not all, insects have
wings. I have one see-through soft
pair of hindwings. They fold under
a hard pair of wings that close up
like a shell. The hard ones are called
forewings or elytra (el-ee-tra).

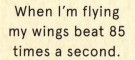

My **elytra** open out
when I need to fly.

My **hindwings**
unfold when
I need them.

When I'm flying
my wings beat 85
times a second.

Many insects have wings that look different to beetle wings.

Butterfly

Dragonfly

Housefly

Lots of insects have different mouthparts to me because of the food they eat.

A leafcutter ant can slice off pieces of leaf.

A butterfly can suck up nectar.

True bugs are a particular group of insects that have a needle-like mouthpart to pierce plants or animals and suck up liquid. (They're the only ones that should really be called bugs, by the way, but we all get the nickname.)

Insects have been on Earth for roughly 400 million years. The largest insect that ever lived was a prehistoric creature, a bit like a dragonfly, with wings that stretched as wide as a crow's.

fossil

Half of all the known animal species on Earth are insects and there are probably lots more still to discover. There are way more insects than humans!

Insects rule!

TEENY-TINY BUG FACT
We ladybirds can smell things with our feet! Lots of other insects can, too.

HOW THE BUTTERFLIES GOT THEIR COLOUR

A bug tale from North America

A magic mixing bag captures summer's beauty

This magical tale is based on a legend of the Tohono O'odham Native American People who live in the Sonoran Desert in Arizona. It's about how butterflies came to be so colourful.

A kind spirit of goodness called Elder Brother watched over the Sonoran Desert People. One beautiful day he sat watching the children play, making sure they came to no harm. The sky was blue, the sunlight was sparkling on the green leaves of the trees, and the perfume of the red and yellow flowers was floating through the air.

However, as Elder Brother thought about the world, he began to feel sad.
"All this beauty will slip away when winter comes," he thought. "The leaves will fall. The flowers will die and the sky will turn from blue to grey. I really must do something to catch the beauty of the things I see today. I'll make something that will bring happiness to everyone."

He took out a magic bag and began to fill it.
First he added a snippet of the blue sky.
Then he popped in a red flower, a yellow
flower and a green leaf. He added lots of
other colours, such as a silvery bird's feather.
Finally, he sprinkled in some sunlight sparkles.

"Come and see what I've made," he called
to the children. They gathered round as he
shook and then opened the magic bag. A cloud
of butterflies fluttered out, in every colour
imaginable. Their beauty made everyone feel
glad and grateful for the world, just as Elder
Brother wanted, and they still do. Remember
that nature's beauty is a wonderful gift you
can see every day.

I know that not all insects look beautiful
to humans – but we're all interesting!
Our bodies are just right for the lives
we lead. Isn't nature incredible?

HOME SWEET HOME
Places we like to live

Insects make their homes in all sorts of different places across the world. You'll find us here, there, and pretty much everywhere!

Surviving in the city

Hiding from the heat

Insects live happily alongside humans even in big busy cities. For instance, around 17 billion ants are estimated to live in the city parks of New York, USA! There's even a bee that digs out the mortar between building bricks to make a tiny nest hole. It's called a mortar bee or masonry bee.

In hot deserts insects usually burrow underground to hide in tunnels during the day. They will pop out at night when it's cooler, to hunt for food. Desert crickets and beetles all hide from the daytime sun.

A frozen habitat

A space home

Antarctica might be the coldest place on Earth, but it's still home to an insect. The Antarctic midge lives in pockets of soil around the edge of the frozen land. It spends most of its life as a little brown wormlike larva. Amazingly it can survive being frozen for up to 6 months.

In 1999 four ladybirds went to live in space onboard a NASA space shuttle. They took part in an experiment to see if they could catch aphids in weightless conditions. It turned out they could! They were nicknamed George, Paul, John and Ringo after the world-famous Beatles pop group.

A TREE FOR ME

Trees make an ideal home for lots of insects. Here's why...

Trees make flowers in springtime. Inside each bloom there is sweet sugary nectar – food for insects such as bees and butterflies.

Leaves are foods for lots of insects such as caterpillars. Some birds eat insects, though, so there are also lots of birds living in trees!

Cracks and crevices in bark make good hiding places where insects can lay their eggs and shelter from the winter weather.

When a fallen tree branch starts to rot the wood goes soft and mushy. It's the perfect food for the larvae of some beetles (such as stag beetles and cockchafers). They hatch from eggs laid on the wood and will tunnel inside the log.

Thousands of mini insects live in soil and lay their eggs in the top layer. The eggs are small and white or pale yellow. They are sometimes laid in clusters (groups) of eggs.

Lots of insects live amongst the rotting leaves that fall from trees and carpet the ground in a forest. It's called leaf litter.

TEENY-TINY BUG FACT

Forty-three different types of ant were found living on just one rainforest tree in Peru, South America.

NIGHTTIME BUGS
Making use of moon and bug light

I go to sleep when it gets dark, but some insects wake up. Meet the moths and fireflies who fly around by the light of the Moon and stars.

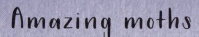

Amazing moths

* Moths come out at night and are known as nocturnal creatures. They work out which way to fly using the position of the Moon and the stars in the sky.

* Moths collect nectar from night-blooming flowers. They can find flowers in the dark because they can smell their scent, and can also detect a gas called carbon dioxide, given off by the blooms. Unlike humans, they can see a type of light called ultraviolet, which is reflected by the flower petals.

* There are roughly 160,000 different kinds of moths compared to just 17,500 types of butterflies.

* Some types of moths have a wingspan wider than this book! The biggest ones are found in tropical forests.

* The clearest difference between moths and butterflies is how they hold their wings when they are resting. Butterflies close up their wings like two praying hands. Moths keep their wings spread. Moths also tend to be much fluffier than butterflies, which helps to keep them warm on chilly evenings.

I think Luna's sleeping in there!

Butterfly

Moth

Luna's hiding place

Fantastic fireflies

✱ Fireflies are a type of beetle. They are also called glow worms or lightning bugs.

✱ The fireflies glow to attract each other, so they can pair up. Then the female will lay eggs. Each type of firefly has its own pattern of light flashes that you can see in the dark.

✱ A male firefly makes its own light using chemicals in its body. It's a clever animal skill called bioluminescence (by-oh-loom-in-es-sents).

✱ The light might be yellow, green or orange. It's not hot like electric light would be, so it doesn't harm the firefly.

✱ One species of female firefly is a deadly faker. They flash the signals of another species, luring the males in so that they can eat them!

At night insects are safe from hungry birds, but there is still danger from bats and owls who enjoy an insect snack.

TEENY-TINY BUG FACT
There are more than 2,000 different types of firefly, but some of them don't glow. Instead, they make special smells to attract each other.

BUG BUILDERS

The best tunnels, mounds and towers

Ladybirds don't make nests but some insects I know are fantastic builders. They make homes to protect themselves and bring up their babies. I think it's time to give out my Bug Builders Awards!

 Best Team

Ants make a great building team. They work together to burrow out tunnels under the soil and make underground ant cities. Some ants also use twigs and leaves to make mounds above their tunnel cities. They live in the mounds in warm weather and hide underground when it's cold.

 Best Gluers

Weaver ants build their nests in tropical trees from southeast Asia down to Australia. Worker gangs of ants roll up leaves and then glue them in place using sticky silk made by their larvae. The ants usually create a nest about the size of a football.

Weaver ants pulling a leaf together to glue it in place.

 **The Champions**

The award winners are the mound-building termites. They live in parts of Africa, South America and Australia. The termites build a tall mud tower called a termitarium. It can reach up to 5 m tall – taller than an adult African elephant.

The queen termite lays eggs inside the nest – one every three seconds. The workers hatch the babies in a nursery chamber.

Worker termites push pellets of chewed soil into the tower, where it dries hard. It takes them four or five years to finish building their hollow tower. After that they must keep repairing it.

Best Chewers

Wasps build nests by scraping up slivers of wood and chewing them to make a sticky pulp. The pulp dries to make a tough paper ball. Inside are six-sided (hexagonal) chambers to hold the wasp eggs. You might see a wasp nest in a sheltered spot such as under the roof of a building.

**A wasp's nest only holds eggs
and hatching wasps, not honey.**

 Best Mini Builders

Caddis fly larvae live underwater in ponds and lakes until they are ready to hatch. To protect themselves from fish and bigger insects, they grab pieces of gravel, sand, twigs and plants, and glue them into a case around their body. It looks like a tiny stony sleeping bag!

Bees are great builders, too. They make waterproof honeycomb using their body wax and use it to store honey as well as larvae. You can find out more about them in another book – *The Secret Life of Bees*.

Termites are farmers. They eat plants but can't digest the toughest stringiest bits. Instead they poo the chewed-up plant material into chambers inside their tower. They add fungus spores and it grows over the plant mush, turning it into crumbly compost that the termites can eat.

Enemies such as aardvarks and driver ants might attack a termitarium, but underneath there are underground escape tunnels for the termites to use.

TEENY-TINY BUG FACT

**Up to two million termites
might live inside a termitarium.**

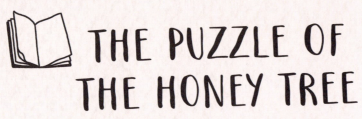

THE PUZZLE OF THE HONEY TREE

An insect tale from southern Europe

A maker mystery solved

This story is one of Aesop's tales. He was an Ancient Greek storyteller from long ago and his tales were often about animals.

Once, there was a big argument between the wasps and the bees from a woodland kingdom. It all began when a waxy honeycomb full of delicious honey was found inside a hollow tree.

"That's ours," said the wasps.
"No, it's not! We made it," cried the bees.

Along came Judge Hornet, who was in charge of the woodland.
"I will decide on this argument. Call the insect court together!" she cried.

The woodland insects gathered under the biggest tree.
The first insect to speak was a moth.
"I saw winged creatures flying in and out of the hollow honey tree. I heard them buzz and I think their bodies were striped yellow and black."

"That was us!" cried the wasps.
"No, it wasn't!" replied the bees.

Nobody could agree or settle the row.
"I need to think carefully about this. Let's delay the court for a few weeks," sighed the judge, but then a wise old bee stepped forward.
"If we wait too long the honey will spoil," she said. "Both the bees and the wasps need to build a waxy honeycomb and fill it with honey. Then we'll see what they both look like and know who made the one in the tree."
"Not fair!" cried the wasps, but the judge was already nodding her agreement.
"That's a great idea. Wasps and bees, start building.
Ready, steady, go!" she ordered.

Of course, the wasps couldn't build a waxy honeycomb, let alone fill it with honey. "The honey belongs to the bees. They deserve it for working hard!" declared Judge Hornet. "You wasps were just trying to benefit from somebody else's efforts. Now buzz off!"

SWIMMING BUGS

Hairy hunters and hatchers

Sit and watch a pond or a river on a warm sunny day and you'll see lots of busy insects. Let's see who's around...

Some insects can swim underwater. They have lots of tiny hairs on their body and a waxy body-coating, too. The wax keeps out water and the hairs trap air, so the insect gets surrounded by its own bubble of air to breathe as it dives underwater.

Insects such as water striders walk across the surface of the water. They have a clever secret – their very hairy legs! The tiny hairs trap air bubbles to help the insects float, rather like someone swimming held up by arm bands filled with air.

Water strider

As well as hungry fish lurking underwater, many pond insects are fearsome hunters themselves. For instance, whirligig beetles swim with their back legs, steer with their middle legs and use their front legs to grab a victim.

Giant water beetles will eat anything they can catch. Some types are nicknamed 'toe biters' because they will even give a human swimmer a nip! The bite won't harm a human, but it is painful.

Whirligig beetle

Giant water beetle

Water bugs such as the giant water beetle have a sharp beak-like mouthpart to pierce unlucky creatures. They inject poisonous spit that turns the prey into mush. Then they suck it up like a nightmare milkshake! Eek!

Lots of insects lay their eggs in freshwater. The eggs hatch into tiny wiggly larvae. They might live underwater for a while before they finally change into adults and fly above the water surface.

Water larvae often look like tiny underwater worms.

When warm weather comes, insects start to hatch. You might see a cloud of them above the water. They won't live for long, though. They will mate, lay eggs and die. Then they might fall into the water and get gobbled up by fish.

Fly fishermen put a fake 'fly' on a hook at the end of a fishing rod. The flies are feathery, with shiny beads, so they look like insects that have hatched from the water. The fishermen hope to fool fish into swimming up and eating the fly, so they get hooked.

TEENY-TINY BUG FACT
Ladybirds can paddle but we would drown underwater. We prefer land!

NOISY BUGS

Clever clicks and brilliant buzzing

Occasionally you might hear us insects. Apart from the buzzing of wings as we fly, some of us make noises to warn off attackers or to attract a mate.

Some insects rub body parts together to make a sound, often to show other insects how big and strong they are. It's called stridulation (strid-you-lay-shun). Grasshoppers rub their hind legs against their wings, while male crickets rub their wings together.

Male cicadas make a buzzing sound by flexing a body part called a tymbal. It's made of tiny ribs that pull and push together. African cicadas are the noisiest. Their sound can be almost as loud as a chainsaw!

CLICK, CLICK!

Some insects are good at mimicking – copying the sound of other insects. Australian katydids can mimic the noise of female cicadas to lure males near. Then... GULP! The male cicadas get eaten!

HISS!

Bats use high-pitched clicking sounds, called echolocation, to find tasty moths. The clicks bounce back from the moths so bats can sense where they are. However, some tiger moths and hawk moths make similar clicking sounds themselves to confuse the bats. Clever, huh?

TEENY-TINY BUG FACT

Ladybirds don't like loud noise. In tests, scientists found we didn't mind soft music but flew away when we heard loud rock!

When they feel threatened, cockroaches hiss by pushing air out through tiny tummy holes. Meanwhile deathwatch beetles create tunnels by gnawing wood and then bump their heads against the walls, making a 'knock, knock' sound to attract a mate.

SQUEAK!

North American hawkmoth caterpillars can whistle! They make loud squeaks through their breathing holes, to scare away hungry birds.

You can hear insects such as bees buzzing because their beating wings are moving air. Bees sometimes buzz louder inside flowers because they are vibrating, or moving, their bodies to shake pollen from the flower. They will take some of the pollen home to feed their larvae.

BUZZ!

Many insects have their own version of ears, often in surprising places such as their knees or tummies. Most beetles don't have ears, though. We 'hear' by sensing vibrations around us.

KNOCK, KNOCK!

THE SONG OF THE CICADA

A bug tale from Ancient Greece

A bug becomes the star of the show

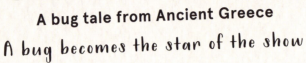

Cicadas live in warm countries, where they nestle in the grass. The loud chirruping sound of the males is hard to miss, so no wonder they have their own story!

Long, long ago in Ancient Greece there was a musician called Eunomos (you-no-mos). He played an instrument called a cithara (sith-ara), which had five strings like a guitar. Eunomos was a great cithara player and singer.

One day Eunomos heard about a music competition, which was to be held at the beautiful outdoor theatre of Delphi. It would be a very grand occasion, as the theatre could sit 5,000 people on its curved stone seats.

"To win, I'll need to practise a lot," he thought to himself, so he began practising day and night until his family and neighbours complained. "Please stop! Your music is good but we're getting sick of it," they begged. "OK. Sorry. I'll go and practise in the empty fields," agreed Eunomos and he took his instrument out to the countryside. The fields were home to the cicadas and they buzzed along as he played and sang.

On the day of the competition there were musicians playing pipes, harps, bells and drums but Eunomos was confident he was the best. He proudly stood at the centre of the stage and began plucking the cithara's strings. Up the notes he went, from the lowest to the highest, singing:

La, la, la, la...

PING!

The highest string broke! What a disaster. But a cicada from the fields had hidden itself in the folds of Eunomos' robe. It hopped out, sat on his cithara, and began to buzz the highest note. Between them they finished the song and – of course – won the competition!

MY GIANT FRIENDS
Who's the biggest?

It's time to meet some of the giants of the bug world. You'd have to go to tropical rainforests, hidden islands or deserts to see them for yourself.

The **Royal Goliath beetle** is one of the world's beetle giants. The males can grow as long as an adult human hand. They live in steamy rainforest areas in southeast Africa.

The males have a Y-shaped horn on their heads, for fighting each other. They battle for food, a mate and a place to live.

Male

Goliath beetles usually eat sugary plant sap and fruit but they might munch on animal dung, too.

6–11 cm long

Males are super-strong and they can lift about 850 times their own body weight. That's rather like you lifting a truck!

Female

They have a pair of soft wings under a set of hard elytra – like mine but much bigger. When they fly, they make a loud thrumming noise.

5–8 cm long

The females have shovel-shaped heads for digging into the soil, where they lay their eggs.

Goliath beetle larvae hatch under the soil. They can weigh up to 100 g, which makes them twice as big as their parents!

There are around 3,000 different kinds of stick insects. The longest one also has a long name - **Phryganistria chinensis**. It's from China and it can grow up to 64 cm long. That's roughly as long as an adult male human arm.

Stick insects feed on leaves.

They usually move around at night. In the daytime they sit still, looking just like sticks!

The world's biggest butterflies are found in tropical lands. The biggest one of all is the beautiful **Queen Alexandra's birdwing** from the Pacific island of Papua New Guinea.

Birdwings feed on the nectar of flowers such as hibiscus blooms.

Females grow the biggest, with a wingspan up to 31 cm (about as long as an ordinary ruler).

The **wētāpunga** is the world's heaviest insect. The females can grow to be heavier than a mouse and about as long as an adult human hand. To see them in the wild you'd have to travel to Little Barrier Island (Te Hauturu-o-Toi) off the coast of New Zealand.

Fossils show that they have been around for roughly 190 million years and lived on Earth at the same time as the dinosaurs.

Wētāpunga mainly feed on fresh leaves, usually at night.

A female might grow to about 10 cm long.

TEENY-TINY BUG FACT
Most ladybirds are between 5 and 8 mm long, so we're not amongst the giants of the insect world!

THE TINIEST BUGS
Fairy-sized and feathery

You'd need a microscope to see the tiniest insects on the planet. They're often smaller than a millimetre. We're big compared to them – but can you spot ten of us ladybirds in this flowerbed?

Fairy wasps live all over the world and most species are smaller than a pinhead. They have tiny feathery wings.

The smallest flying insect in the world is a fairy wasp called *kikiki huna*. Its name is Hawaiian and means *tiny bit*. It grows roughly 0.15 mm long (about the same measurement as two human hairs side by side).

The tiniest flightless insect is a male fairy wasp that's only 0.13 mm long. It's found in the US and it has a very long name – *Dicopomorpha echmepterygis*!

Fairy wasps aren't cute like storybook fairies, I'm afraid. They lay their eggs inside the eggs of other insects and the baby fairy wasps eat the lot.

Featherwing beetles are the smallest beetles of all. Most of them are less than 1 mm long. Their wings are bristly fringes that move in an unusual figure-of-eight shape as they fly. They live amongst fallen leaves and logs, where they eat fungi.

The tiniest butterfly is the **pygmy blue**, found in North America. Its wingspan can be as small as 12 mm, roughly the size of an adult human's little fingernail.

Mosquitos are small but nasty bloodsucking insects that can spread disease. The smallest one is the pale-footed uranotaenia from the southern USA. It measures 2.5 mm long but doesn't drink human blood. It prefers to feed on frogs.

TEENY-TINY BUG FACT
The smallest ladybird species is 2–3 mm long, roughly the size of two pinheads side by side.

WHAT WE SEE
Our brilliant bug eyes

Insects have eyes that look very different to human eyes. This means we see the world differently to you.

Humans have two eyeballs, and each one contains a lens that brings you a picture of what you see. Your brain mixes the two pictures to make one. But insects have loads of lenses, sometimes thousands! Our eyes are called compound eyes.

Some of us have eyes on top of our heads and can see right round behind us (dragonflies, for example). That's good for spotting prey and for dodging enemies.

Some termites and insect larvae can see very little, and some have no eyes at all.

Ladybird eyes and antennae, close-up.

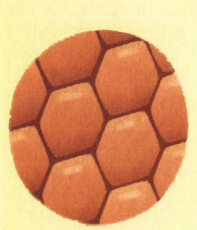

Each insect eye is made of tiny tubes called ommatidia. They are hexagonal, which means they have six sides, and each one has its own lens at the front. So if you look at an insect eye it looks like a mosaic of tiny tiles. We don't see lots of mini images side by side, though. Like you, we mix them together and only see one big picture.

Many insects can see ultraviolet light, a type of light invisible to humans. It gets reflected off objects such as flowers, helping insects to spot them.

Who sees what?

Insects have different numbers of ommatidia in their eyes. The ones with the most ommatidia see the most detail.

Butterflies and moths have around 17,000 ommatidia per eye. They are very short-sighted, but they can see a much wider picture than humans.

Some dragonflies have 30,000 or more ommatidia per eye. They're great at spotting movement, which is good for hunting. Some dragonflies have a darker part on top of their eyes. This may act like a pair of sunglasses to shield them from strong sunlight.

Honey bees have about 5,500 ommatidia per eye. They are really good at seeing purple, violet and blue and they're much better than humans at seeing things when travelling fast.

Luna's eyesight test

What's my ladybird eyesight like? If I were to ever go to an insect optician, here's what the report would say. (It would be different for other insect species.)

- Luna can only see in black, white and grey.

- Her eyesight is pretty blurry. She's very short-sighted and only really sees things close-up.

- She is attracted to light-coloured flowers because they're the ones she sees best.

- She can't see in the dark at all.

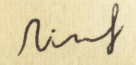

TEENY-TINY BUG FACT
Caterpillars can barely see at all and rely on sensing or feeling things around them instead.

MOST AMAZING
Incredible insect superpowers

If you are surprised by how we see, get ready to be truly amazed by some of the other talents we have!

TEENY-TINY BUG FACT
Ladybirds sometimes fly up to 1118 m high – higher than hot air balloons. We also hitch rides on planes, cars, boats and trains!

THE BEST AIR ACROBAT
Dragonflies have bodies and wings shaped rather like aeroplanes, and like an acrobatic plane they can do some stunning tricks. They catch flies with their feet in mid-air and can eat hundreds of mosquitos a day.

Zoom

THE HIGHEST FLYER
Bumblebees have been spotted on mountainsides at heights of 5,600 m. Tests suggest they could fly even higher.

Buzz... buzz

THE BEST JUMPER
The **Little Froghopper** holds the record with a 70 cm jump, though it's only 6 mm long. That's a bit like a man jumping over a 72-storey skyscraper in one leap. Froghoppers usually live in amongst wild plants in Europe, North America and Africa.

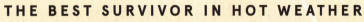

Ready...

steady...

THE BEST SURVIVOR IN HOT WEATHER
The **Saharan silver ant** can go out in heat that would instantly shrivel other insects in its African desert home. It waits until all its enemies (such as hungry lizards) have hidden from the searing midday sunshine. Then it runs over burning sand on its long legs.

Sizzle... sizzle...

... spin!

... loop the loop

THEIR SECRET 🔍

A dragonfly can fly backwards, spin in mid-air, fly upside down or hover like a helicopter. That's because its wings can move separately to each other – up and down or backwards and forwards. It can adjust them to move in lots of different ways.

... fly up high!

THEIR SECRET 🔍

Bumblebees are good at flying and sometimes even nesting in cold high locations because they have such thick fuzzy coats. Their hair is longer than other bees.

boing!

THEIR SECRET 🔍

The froghopper's legs have very strong muscles. They contract, or squeeze, and lock in place, a bit like a string pulled back on a bow and arrow. Then... boing, the froghopper takes off in a millisecond.

run, little ant!

THEIR SECRET 🔍

The top and sides of this ant's body are covered in tiny silvery hairs which reflect hot sunlight away, so it's just like it's wearing a coat of mirrors!

HOW THE RACE WAS WON

A story from Brazil

A coat of many colours

Beetles might not get the insect award for air acrobatics, but we are often amongst the most colourful of the bugs. Here is a folk story from Brazil imagining how a rainbow-coloured beetle came to be.

Long ago all jungle beetles were dull-coloured. Not for them the flashy feathers of the jungle birds or the bright stripes and spots of the snakes and the butterflies. One day, a small brown beetle was slowly crawling along a fallen log on the jungle floor. She was inching forward, minding her own business, when a rat popped out.

"Look at you. You're so slow! You'll never get anywhere. Watch how it should be done," cried Rat. Then he began running along the top of the log and back again. "You could never beat me. I'm as speedy as can be!" he crowed.

A parrot had been listening and swooped down from the trees above.

"Why don't you two have a race?" she suggested. "I'll get the jungle tailor bird to make the winner a bright new coat, in any colour you want."

"Yes, please. I'll have it gold and black like jaguar fur," cried Rat.

"I sometimes dream of having a shiny rainbow coat," sighed the beetle.

"Dream on, slowcoach," scoffed Rat.

The two animals lined up for the race and the parrot gave the starting signal.

"Squawk! Go!"

At first Rat ran as fast as he could and left the beetle far behind. Then he began to slow down.

"This is so easy I may as well just walk. There's no need to get out of breath," he thought to himself.

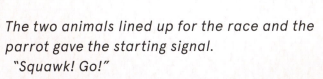

FIN ISH

However, when Rat reached the finish line, he got a very big surprise indeed. The beetle was already there, as calm as you like.

"Hey! How did you do that?" Rat spluttered. That's when the little beetle showed her secret by unfolding her hidden wings.

"Nobody said we had to run, so I flew," she explained.

"I didn't even know you had wings!" Rat gasped.

"Never judge anyone by looks alone," chuckled the parrot. Then she arranged for the beetle to get her prize – a shimmering rainbow-coloured coat that many jungle beetles still wear today.

THE GRUESOME GANG
Grabbers, sprayers and chasers...Eek!

Some of us insects are fearsome hunters with special tricks and weaponry as deadly as any cartoon supervillain. Meet an especially gruesome gang. Luckily they are all way smaller than humans!

PRAYING MANTIS

Home:
Worldwide

Weaponry:
Forelegs armed with hooks and spikes, for grabbing food.

Prey:
Insects and lizards.

Top hunting skill:
Speed. It sits completely still, waiting for prey to come near. Then it ambushes them incredibly quickly.

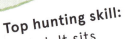

CREMATOGASTER ANT

Home:
Africa

Weaponry:
A spray of toxic venom that paralyses its victims.

Top hunting skill:
It raises up its sting like a hosepipe and shoots poison into the air. It can aim the sting in any direction.

Prey:
Termites, mainly.

ASSASSIN BUG

Home:
Worldwide

Weaponry:
Deadly spit that kills instantly and turns the insides of the prey to liquid.

Prey:
Bees, wasps, dragonflies, beetles and other flies.

Top hunting skill:
Perches somewhere and waits for a victim to fly by. Then it gives chase, bites its prey and injects poisonous spit.

ANTLION LARVAE

Home:
Worldwide

Weaponry:
Big cutting jaws, pincers and venom.

Top hunting skill:
Digs a funnel-shaped pit hidden amongst leaf litter, for passing ants to fall into.

Prey:
Mostly juicy ants.

ARMY ANT

Home:
Tropical parts of the world

Weaponry:
A sting plus sharp jaws like a pair of scissors.

Top hunting skill:
Working together. A swarm of thousands might mount a raid, streaming out of their nest in columns to grab whatever they can find.

Prey:
Any small animals in their path.

TIGER BEETLE

Home:
Different species worldwide

Prey:
Aphids, caterpillars, all sorts of small insects and even slugs. It eats its own bodyweight in food every day.

Top hunting skill:
Running across the ground on its long legs. It's one of the fastest bug runners.

Weaponry:
Sword-like harp jaws.

TEENY-TINY BUG FACT
Insects that hunt actually help humans by eating pests such as mosquitos and greenfly.

FIGHTBACK!

How to beat the gruesome gang

Though we might have enemies bigger and fiercer than us, we insects sometimes have surprising defences of our own. It's time to let you know some top-secret fightback skills!

ACID

If it's threatened, the **red wood ant** can spray out acid that's powerful enough to eat through its enemy's skin.

PINCERS AND POISON PAINT

Soldier **termites** have long pincers to defend their nest from invaders. Some even have an extra-long nose which they use like a paintbrush to smear poison on their enemies.

SUPER-HOT POISON

When **bombardier beetles** are threatened, they fire out an explosive stream of poisonous chemicals, as hot as boiling water. Toads like eating beetles but they'll quickly spit one of these yucky bombardiers out!

PATTERNS TO FOOL YOU

Butterflies and moths sometimes have eye spots on their outstretched wings, thought to help scare off predators. It's possible they might startle an enemy into thinking the butterfly or moth is a bigger creature, perhaps a bird or even an owl.

DON'T MESS WITH LADYBIRDS!

Red, yellow and black are a kind of danger sign for creatures. They send a message – THIS CRITTER MIGHT BE POISONOUS. Our bright ladybird colours are a warning to animals that we might taste bad, and we do!

I remember these spotted red bugs taste yucky.

Our colours and spot patterns make us easier for birds to remember. They won't forget that we taste bad!

We squeeze out a stinky chemical mixture called pyrazine through our leg joints if we are threatened.

We also defend ourselves by 'playing dead' – drawing our legs in and sitting absolutely still.

Harlequin ladybirds are a problem for other types of ladybirds (though not for humans). They are spreading around the world and they have an extra-big appetite. Not only do they take all the aphid food, they also eat our eggs and larvae. They have similar colours and spots to the rest of us, but they have orange tummies and legs, while most of us have black tummies and legs.

TEENY-TINY BUG FACT
Ladybirds might bite if threatened, though we can't nip through human skin.

ACTING STARS
The art of clever camouflage

Some bugs are masters of disguise, as good at acting as any film star. They might use their disguise to hunt, or to hide from enemies.

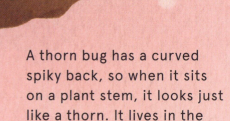

The North American lunar moth looks exactly like a leaf when it sits still with its green wings open.

A thorn bug has a curved spiky back, so when it sits on a plant stem, it looks just like a thorn. It lives in the Americas and parts of Asia.

No wonder the dead leaf mantis lurks amongst fallen leaves in its Malaysian home. Its prey won't spot this hungry enemy until it's too late!

The flat bark bug looks just like a knobbly section of tree bark, the perfect woodland disguise. It's found in most parts of the world.

BRAVO!

SUPERB!

A sand grasshopper is coloured just like grains of sand. It's really hard to spot on the North American prairies, where it hides out in sandy locations.

The orchid mantis pretends to be a pretty bloom full of yummy pollen and nectar for visiting insects in South Asian forests. In fact, those insects end up being the meal!

We've heard about stick insects already (see page 27) but their southeast Asian cousins the walking leaf insects are pretty amazing, too. They have a long body that looks like a branch of leaves. They grow about as long as the palm of an adult human hand.

Peppered moths are speckled, as if they've been showered with pepper from a pepper pot. It makes them super-hard to see against pale-coloured tree trunks covered in lichen. They live in Europe and North America.

WONDERFUL WORK!

TEENY-TINY BUG FACT
The orchid mantis can change colour to blend in with its background.

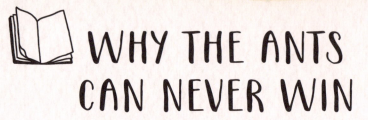

WHY THE ANTS CAN NEVER WIN

A story from South Africa

A wise king shows the way

Life in the wild is tough: big animals eat little animals and that's what this story is about. It's based on a folk tale from South Africa.

The ants were fed up with having so many enemies. Aardvark liked to slurp them up with his long tongue and Lizard grabbed them for a snack whenever he could. Then there were all the hungry birds. The danger was never-ending, and the ants wanted to make it stop.

The different species of ants held a meeting to find the answer.
"We should build safe dens underground," some cried.
"No! Let's build nests above ground," shouted another group.
"That won't do. We need to live up in the trees," another group insisted.

The ants couldn't agree so they all went back to their homes and did what they thought best. Some of the ants built a den underground.
"This is safe from big old Aardvark and those greedy birds," they said, but it turned out that Lizard could wriggle in and reach a few of them with his flicking tongue.

Some of the ants built a den above ground.

"This is too strong for Lizard or the pesky birds," they agreed, but it turned out that Aardvark could break in with his sharp claws and lick up an ant meal or two.

Some of the ants climbed up to live in tree holes. They were safe from Aardvark up there, but Lizard could climb along the branches and it was perfect for the birds.

"We failed! We should have found a way to work together," the ants cried, and began to blame each other. Eventually the wise King of the Insects (the biggest beetle in the land) called the ant leaders to his palace hidden under the bushes.

"Working together is a good thing," he explained, "but every creature in the world can't do that. It's impossible. My advice for everyone – bugs big and small – is this: do your best with the world you have."

The Insect King was right. You can't stop wild creatures eating each other because they're all connected. Some eat plants. Others eat the plant-eaters and each other. The connections are called a food chain or a food web.

THE WORLD NEEDS BUGS!
An insect SOS

Insects are vital to our planet because we help plants grow and we eat up lots of the pests that harm humans and crops. Some of us are even helpful planet cleaners – munching up waste such as dead leaves and even dead creatures. But we're in real trouble. Our numbers are falling, and we need your help.

Wind turbines turn wind into power.

Humans need to find ways to reduce harmful pollution, such as that caused by burning coal or oil. Using power from the sun and wind is much kinder to the planet.

Solar panels turn sunlight into power.

CAFE

WE USE ORGANIC INGREDIENTS

Organic means no harmful chemicals were used to help grow the ingredients.

Vegetable garden

Harmful chemicals used in farming are a big reason bugs are dying out. You humans can help by pushing for farms to use fewer and safer chemicals. Also, if you have room, try growing your own vegetables without using chemicals.

A wildflower meadow gives us lots of food.

Ninety per cent of flowering plants rely on insects visiting them to feed on nectar. Pollen is dusted onto the insect when it visits and gets brushed off when it lands on a different flower.

The pollen then joins with egg cells inside the flower to make seeds. It's called pollination, and the more pollinating insects – such as bees and moths – there are around, the better.

Parks and green spaces in towns and cities are good for us as they tend to have lots of flowers and ponds, too. Cutting the grass a lot and using weedkiller destroys the wildflowers many of us rely on, so it needs to be avoided.

A pond makes a great home for insects if it's looked after and kept clean.

Lots of bright white outdoor lighting is harmful to nighttime insects because it attracts so many bugs. It confuses them! Amber or yellow-coloured LED light or warm white solar-powered light is better for bugs.

TEENY-TINY BUG FACT
Seventy-five per cent of the crops that humans grow rely on insects to pollinate them. You definitely need us!

BE A LADYBIRD FRIEND
Make a cosy flowery home

There are lots of things you can do to help all the friendly little bugs who live near you. In return some of us will eat up the pesky aphids who attack your plants, while some of us will pollinate your flowers for you.

MAKE A LADYBIRD LODGE

Build a lodge to help ladybirds sleep through the winter in a safe place. Position it in a sheltered spot that won't get too hot or cold, for example against a wall or a shed.

You will need:

· Some dry pine cones
· Dry twigs and leaves
· Any hollow twigs you find

· Two spare roof tiles or two scrap pieces of wood to make the roof

Pile the pine cones, twigs and sticks on the ground. Then lay the tiles or pieces of scrap wood over them like a house roof, to keep them dry. Fill in any gaps with more dry leaves.

If you don't have tiles or wood for the roof you could stuff your cones and sticks into an old flowerpot or two, and lay them sideways. Use stones to wedge them in a safe spot.

PLANT FOR US

Even if you don't have a garden, ladybirds might visit if you grow plants in pots on your windowsill or doorstep. Here are some flowering plants that attract ladybirds looking for aphids.

Dill

Fennel

Calendula

Sweet alyssum

Marigold

HELP US STAY OUTDOORS

Ladybirds sometimes come indoors for our winter sleep, but it's not good for us. If the heating comes on, we could wake up at the wrong time, when there are no aphids to eat. If you find a ladybird indoors in winter, gently take it outside and put it in a hidden sheltered place, safe from harm.

Too many aphids will damage a plant
but we can help by munching them up!
We'll spread smiles when we visit you
and fill our tiny tummies, too.

And when you see a ladybird
stretching out its little wings,
ready to take off and go...
tell them Luna says hello!

Luna the ladybird

47

For all the bugs
and butterflies.
We need you! - M.B.

To Darlie, who loved laying in
fresh spring grass cuddling
the butterflies - V.M.